To Mom & Dad

To all my staff, for their unqualified support

And to my clients, for letting Supon Design Group express its talents and creativity through their projects.

© 1995 by **SDPress** c/o Supon Design Group,
International Book Division

All rights reserved. No part of this book may be reproduced in any form without written permission of the copyright owners. All images in this book have been reproduced with the knowledge and prior consent of the artists concerned and no responsibility is accepted by producer, publisher or printer for any infringement of copyright or otherwise, arising from the contents of this publication. Every effort has been made to ensure that credits accurately comply with information supplied.

First published by:
SDPress
c/o Supon Design Group, Inc.,
International Book Division
1700 K Street, NW, Suite 400
Washington, D.C. 20006
Telephone: (202) 822-6540
Fax: (202) 822-6541

Distributed to the book trade and
art trade worldwide by:
Rockport Publishers
146 Granite Street
Rockport, Massachusetts 01966
Telephone: (508) 546-9590
Fax: (508) 546-7141

ISBN 0-9644038-0-3

10   9   8   7   6   5   4   3   2   1

Printed in Hong Kong

DESIGN DIRECTION

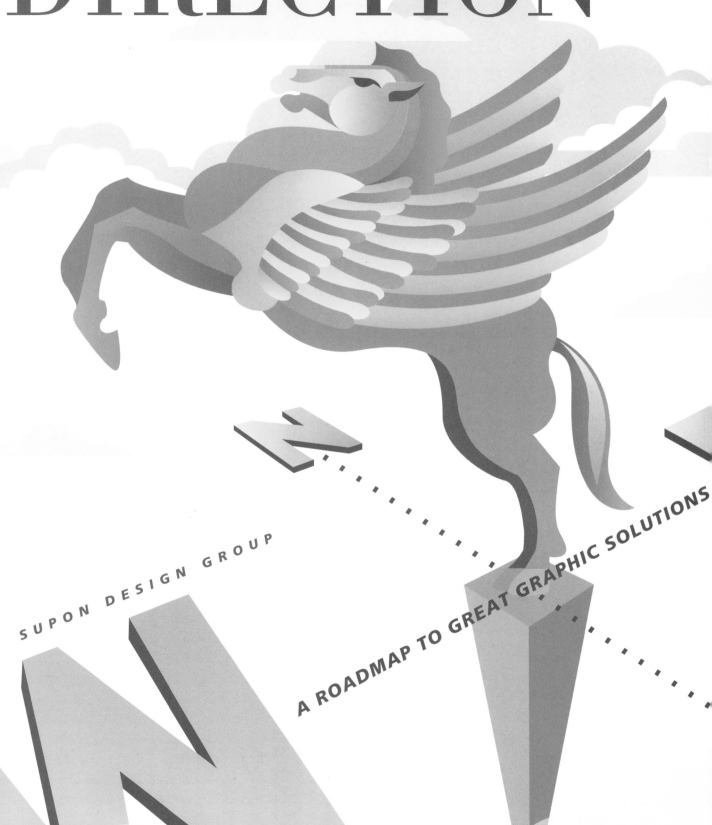

SUPON DESIGN GROUP

A ROADMAP TO GREAT GRAPHIC SOLUTIONS

Creative Director & Managing Editor
Supon Phornirunlit

Art Director
Andrew Dolan

Jacket Designer
David K. Carroll

Book Designers
David K. Carroll, Supon Phornirunlit

Writers
Linda Klinger, Wayne Kurie

Editor
Wayne Kurie

Support Staff
Andrew J. Berman, Jacques Coughlin, Stephen Delin,
Mimi Eanes, Anthony Michael Fletcher, Michael LaManna,
Usha A. Rindani, Apisak "Eddie" Saibua, Deborah N. Savitt

Camera Services
CompuPrint, Washington, D.C.; Color Imaging Center

Photographers
Barry Myers, Oi Veerasarn

The following individuals are acknowledged for their assistance in making this project possible: Henry Kornman, Andrew Clarke, Winnie Danenbarger, and G. Stanley Patey. Sincere appreciation must also be given to Supon Design Group's clients, without whose continuous support our company could not exist.

The text in this book is set in Adobe Frutiger and Bodoni.

TABLE OF CONTENTS

ABOUT SUPON DESIGN GROUP   5

INTRODUCTION   6

PORTFOLIO

GRAPHICALLY: ELEGANT   10

GRAPHICALLY: CONTEMPORARY   28

GRAPHICALLY: PLAYFUL   60

GRAPHICALLY: CORPORATE   82

GRAPHICALLY: PERSONALIZED   100

ABOUT SUPON DESIGN GROUP

As a team of business people who design, write, and illustrate, Supon Design Group's rapid growth is attributed to the practice of its belief that good design incorporates good business, fair prices, commitment, and excellent client rapport.

A varied cross-section of clients too numerous to list, from tiny entrepreneurships and modest, non-profit organizations to multinational conglomerates, ensure projects remain varied in concept and intent. With their client focus continuing to expand to include international interests, Supon Design Group nonetheless maintains its policy of close communications with all clients, encouraging those geographically located to visit their spacious studio, converse with the designers, and review their work in progress. Staff members are experienced in creatively realizing a wide assortment of design tasks, from the subtle nuances of a finely cast corporate identity, a lively magazine layout, or attention-grabbing 3-D packaging, to campaign components that remain attractive and consistent from poster to premium.

Each project offers unique challenges, but each also demands vision and an ability to accurately illustrate the client's objectives in graphics and symbols and type.

Today, the Washington, D.C.-based organization consists of three divisions: the graphic design studio; the international book division, credited with publishing numerous full-color volumes on a variety of industry topics; and the product division, whose innovative creations include software for graphic applications and innovative paper products. The Supon Design Group studio eschews stereotypical definitions of design services; it creates original concepts according to client need and refines those concepts based on client interaction, purpose, and philosophy. The designers have become well-known for their production of marketable concepts and successful visuals because of their consistent determination to strive for the exceptional.

Supon Design Group's award list continues to grow, counting among its hundreds of honors every major design award in the nation, including Gold Awards from the Broadcast Designers' Association and the American Institute of Graphic Art, Baltimore Chapter; Silver Awards from the Direct Marketing Association of Washington; and Addy Awards.

Marketing is a priority for the firm, which has won numerous awards for its own inspired, eye-catching promotions. The studio and its design have appeared in such well-known industry publications as *Print*, *Communication Arts*, *Step-by-Step*, *HOW*, and *Graphis*, as well as Asia's *Media Delite*, and Germany's *Page* magazine. Because its work is recognized around the world, Supon Design Group has developed an enviable reputation in the competitive industry by positioning itself as a studio on the cutting edge of design, but firmly grounded in principles of commitment, hard work, and a partnership with clients — principles that continue to produce outstanding yet thoughtful design solutions for a variety of objectives.

INTRODUCTION

Oliver Wendell Holmes wasn't speaking graphically when he said, "The human mind, once stretched to a new idea, never goes back to its original dimensions."

But he could have been. Creativity — the foundation of our industry — is just a culmination of new ideas, brought to a visual reality — the imprecise flourish that tops off the graphic project and makes it sensational. And the more the mind is used to generate new ideas, the more one strengthens his or her creative tendencies. The minds of designers must be stretched to extraordinarily broad dimensions with the outpouring of original concepts that is part of their daily routine.

Creativity on its own, however, is too nebulous to be successfully applied. Like an unruly offspring, it needs boundaries and discipline. That's where direction comes in. Direction in design is like a roadmap of procedures. It is the fundamental strategy — the motivation behind the piece. Direction gives a clear purpose to the myriad of images that eventually fuse to make the solution sparkle.

GUIDING THE GRAPHICS

Direction lends the proper symmetry to communication and aesthetics. Infuse ideas with information (creativity), structure it with a plan of attack (direction), and the result is graphics that reveal a union of concept, communication, and image.

Design direction, however, is an elusive concept. Like "beauty" or "quality," it's better defined by example, or realized in some type of application. Design without direction, for example, is just a group of elements, puzzle pieces, a meandering combination of color and type. Direction guides the solution to completion, providing a way for an idea to be successfully defined. It keeps the concepts from getting lost in a maze of attempts and experimentation.

There is a symbiotic relationship between direction and creativity. Creativity, when linked with style, is the "dress" of the project — how well the shirt flatters the pants, the sauce complements the steak, and the concept matches the client. With its emphasis on proper organization and purposeful arrangement, an early direction will ensure the piece positively captures a corporate characterization. Yet without a balance of each, and without creativity coming into play every step of the way, the piece won't have the strength to carry a concept to completion.

Some designers approach the design process metaphorically. But that approach won't yield satisfactory results if the solution doesn't "fit" the project. That means first addressing the targeted company's intention, its personality, and its reputation, directing the whole result to the right audience, and then building the visuals from that foundation. In the end, it is the design's literal application as well as its incorporated analogies and metaphors that will ensure the piece has success and longevity. A common design tendency may help studios develop an identifiable style, but each project must activate its own original proclivity.

WHERE WE FIT IN

At Supon Design Group, each project's direction incorporates basic tenets of our philosophy. To understand our studio within the business environment, one needs to understand the three major components of this philosophy: integrity, marketing emphasis, and vision. Together, they represent the interaction of visual appeal with elements of business acumen. Singularly, marketing emphasis is most important. A solution that sells the product or service is our common goal with each client. We strive to give clients what they need, instead of what they want — and the critical difference is that, because need is always based on close interaction and good communications, customer satisfaction is ensured.

Getting from point A to point B requires a designer who concentrates on the formulation of the project's meaning rather than its illustration. Creating the right look embodies objective and character, and incorporates a thought process that goes beyond an easy solution — a pretty picture — and responds to the real concern — increased sales, or a new posturing in the marketplace, or an introductory promotion for a new company. We must create a piece that not only looks good, but persuades, educates, excites, unifies, and encourages its audience, while also making a clear marketing statement to the public.

Vision, on the other hand, is often defined as giving shape to the invisible. Vision ensures that everything about a successful solution fits the intent, from its fundamental idea to the choice of technology or medium to produce it. The solution fits because the designer not only embodied the concept, but determined the right direction.

And our philosophy is also based on integrity — the fact that we use our skills foremost to benefit the client. Design cannot be created for the sole purpose of winning awards; that motivation can drown designers in their own ideas. There's nothing wrong with educating the client about new perspectives, but his or her interest should never be overlooked. We believe ideas should be flexible, not restricted by designer ego. One must listen closely to the client's vision of how to achieve his or her goals, and use talent and experience to expand that vision, not destroy it. But it takes discipline to avoid being seduced by the sheer visual beauty of what designers do or the limitless capacity of our technology. The direction has purpose, and should not stray too far into the art or mechanics of design and forget about the bottom-line industry within which design is firmly ensconced, and to which the solution must ultimately respond.

NOT JUST A
PRETTY PICTURE
At Supon Design Group, we integrated direction as the method added to the philosophy upon which our company is founded, making direction an important component to design success. When clients approach us with a somewhat limited vision of their message, our creativity provides the ability to stretch that vision — push the envelope, as it were — to get something practical, appropriate, but nonetheless eye-catching and with marketing savvy. The fundamental direction, then, is an element inseparable from the business aspect — the rules of the game. Direction is the precise strategy that contrasts with creativity's cryptic imprecision.

The roots of the creative process are frustratingly enigmatic. They begin with a need, then an idea. Direction doesn't exist in a vacuum, but is given substance through detailed client/designer discussions, an understanding of goals, and — more important than these — the singular ability to interpret intentions accurately. That, of course, falls under the auspices of basic business sense.

With the added support of solid business components and skill, the ultimate solution will fit because metaphors work — they get across the right proportion of humor and hard sell, philosophy and marketing finesse. The style fits because it espouses images that distinguish the client — combinations of uniqueness and lyricism that have never been presented before in exactly the same way. The project's success is more likely because it emphasizes the right elements with color and typography, rule and proportion and grid.

In the following pages, we've presented some examples of what we've learned about mapping out great solutions. Each graphic bearing results in a solution defined by its original cast and perspective. Page through these visual solutions, and we think you'll find ideas that not only sell, but are built on the solid groundwork of effective communications.

GRAPHICALLY ELEGANT

FEATURING ELEMENTS THAT COULD BE CALLED SOPHISTICATED OR EVEN AUSTERE, THE ELEGANT STYLE CUTS THROUGH THE SUPERFLUOUS COMPONENTS TO THE HEART OF THE SOLUTION. Colors, which act more like subtle accents than bold couriers of theme or statement, favor the quieter tones — the blues, the golds, the creams. Artistic renderings often work well in elegant design, in place of boisterous, dynamic pictures that shout for your attention across a room. Elegant design finds the underlying metaphors and builds interest using a light and gentle hand, a studied approach, and usually a steady eye that can see through the banal and sculpt an original approach each time — slight, and with fine distinction. There is an element of succinctness to this style, and companies from clothing manufacturers to financial conglomerates, when asked for their perception of their company's direction, often find their vision is made clear when adapted to an elegant style.

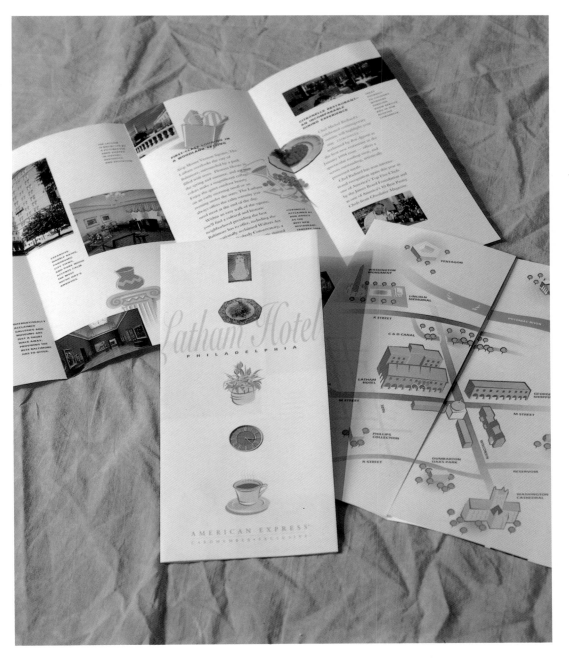

HOTEL DIRECT MARKETING PROJECT

Targeted to Washington, D.C., Philadelphia, and New York business travellers who do not stay at the Latham Hotels, this brochure, letter, and envelope stress, among other features, the benefits of the neighborhoods in which the hotels are located. Design and photos emphasize the dignity of the hotels, whose promotion capitalized on their personalized service. The typeface was selected for its accessibility, with Latham green the predominant shade, and gold added for a touch of the majestic.

MUSEUM OF JUNK

PETER LEUNG COMPANY

THE MUSEUM OF JUNK CORPORATE IDENTITY

We chose the name of Peter Leung Company's department store, whose product displays have a distinctive, museum-like flair. Then we not only refined the concept for this highly original retail venture, but also developed both the logo and a complete presentation in a matter of weeks.

A place where an original cast-iron piggy bank may sell for $4,000 and an imitation for $39.95, the Museum of Junk wanted an identity that would embody both its refined and whimsical natures. Its respect for the environment is suggested by muted Earth tones.

The elements of the logo represented the classic (the trash can/column metamorphosis), aesthetic (the picture frame), and temporal and memorable (the clock).

The transcontinental project owed its successful completion to the ability of designers and client — who had never met — to work together.

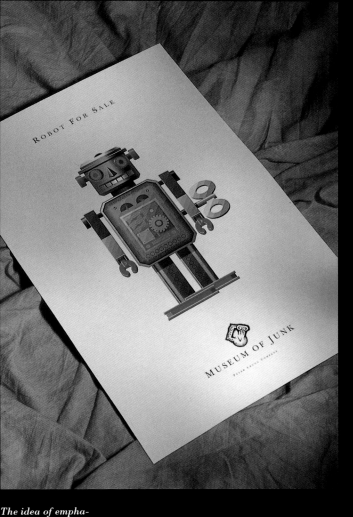
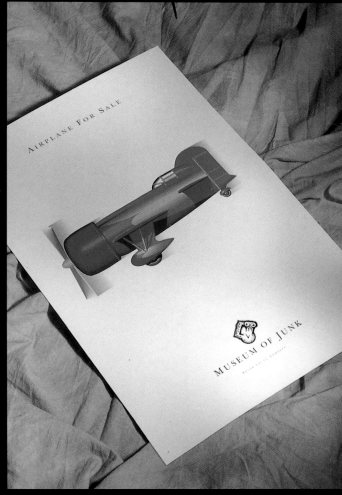

The idea of emphasizing the "trash can" image was quickly abandoned in favor of visuals of nostalgia and memory-makers — the type of articles found in an attic that are called "junk" by some and "treasure" by others.

When designers drew the frame surrounding the logo, it provided an entirely new and interesting element for the T-shirt application.

Standard Kraft paper was chosen for the shopping bag to underscore the antique theme, and both the bag and tag were designed at the same time to ensure similarity.

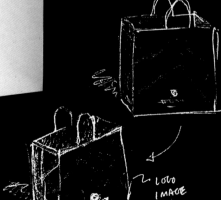

FRESH MARKET LABEL DESIGN

To develop a new label design for a container that housed cut, ready-to-cook produce, a stylized graphic was used that made the Fresh Market product appear even more upscale and appealing. The classic, softer lines were intended to appeal to urbane consumers — the target market of the product — and the colors contrasted well against the white background to catch the eye.

COMMEMORATIVE ANNIVERSARY PROMOTION

In honor of its 25th anniversary, The George Meany Center for Labor Studies tasked us to produce celebratory but businesslike poster, banner, and T-shirt designs. The poster's dominant symbol, an abstraction of the campus's preeminent structure, symbolizes the Center and was adapted for the T-shirt. Its classic yet progressive approach is reinforced by vivid, earthy autumn tones. Refined typography connotes the Center's intellectual underpinnings.

MIDWINTER
PROGRAM INVITATION

The midwinter program for the Food Marketing Institute was held in warm, sunny Florida. To emphasize the welcome contrast in temperature, the organization needed an invitation whose colors complemented the southern hotel's classic tropical color scheme, as well as appealed to the CEO of the institute, who preferred an elegant tone for the publication.

**WEARHOUSE
PACKAGE DESIGN**

To classically communicate a sports motif for Wearhouse, a maker of athletic clothing, the design concentrated on customized illustrations of the three main sports activities for which the clothing is designed — racquet, all-purpose, and track and field.

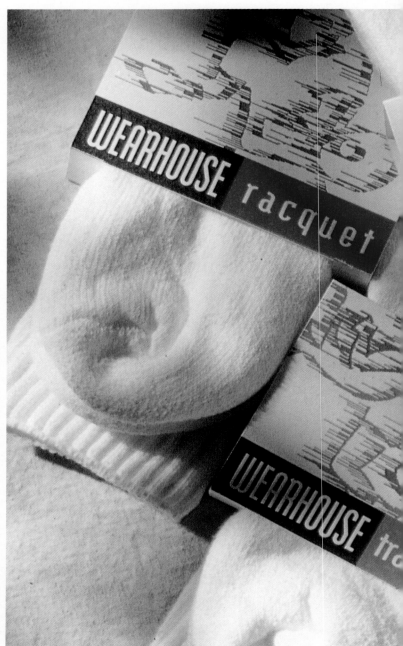

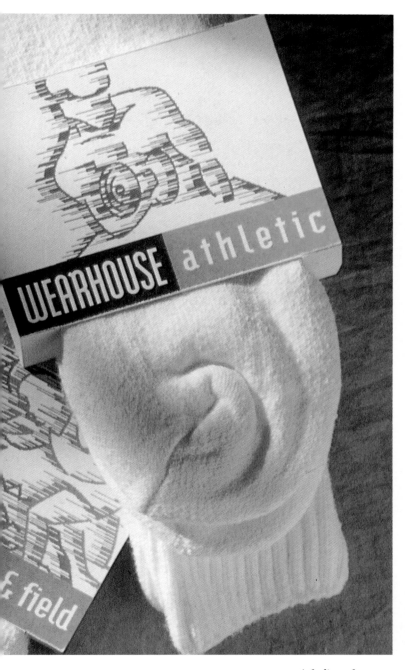

A feeling of naturalness was conveyed with recycled paper and nuances of color, and horizontal lines conveyed movement and energy.

GREAT GATSBY CELEBRATION INVITATION

We were asked to design an invitation, table tent, and other items for a party held by ISL representatives for selected World Cup Soccer supporters. With a "Great Gatsby" theme, the deco motif was an obvious choice, and it also mirrored the interior design of the party's location. In addition, cool colors, especially white, provided a tie-in to the soccer industry. Because of the exclusivity of the guest list, a regal look was incorporated.

"NORMANDY THE GREAT CRUSADE" PROMOTION

To honor the heroes of Normandy and call attention to The Discovery Channel program of the same name, we developed a campaign using the existing program identity. Colors complemented the "Normandy" logo, and "old" photos reminded viewers of the famous allied invasion years ago. The promotion highlighted people in various communities whose names were submitted by viewers and whose selfless acts could be considered heroic.

GRAPHICALLY
CONTEMPORARY

WHY WOULD ONE DESIGN WITH A CONTEMPORARY STYLE? Because the project is looking for a revolution, a facelift, or a dose of energy. Contemporary projects can even be called trendy and be successful, if their lifespan is short, such as an invitation. Progressive work, however, usually better fits pieces that need the durability of a carefully orchestrated design to help them stay effective through changing seasons, new appointments in management, or unpredictable public taste. This modernistic style may be dated by its decade in the future, but not in a way that will make it old-fashioned, stale, or betray its age; rather, contemporary design that is well conceived can reflect the best of its time, and render it timeless. We've seen color contribute to or strengthen an exotic look, as well as bold typography, cutting-edge computer techniques, or a jaunt with experimentation in contour and photos, patterns and shapes. This is a style that may add a personality trait to an image, or strong emotion to an illustration, or even provide a complex story behind a simple name.

U.S. POSTAGE STAMP RECEPTION INVITATION

To introduce a new postage stamp honoring the 1994 games, we developed this die-cut, soccer ball-shaped pocket folder for the U.S. Postal Service reception. Indistinct images ensure countries are not recognizable and speak to all participants of the games. The piece contains a collector's set of stamps with the first day cancellation imprint, guaranteeing its status as a collector's item. The design style telegraphs movement and complements the postage stamp illustration.

COMMUNICATIONS STRATEGY BROCHURE

Our communications strategy was directed to the organization behind World Cup Soccer and built on a striking presentation: a wooden box that held a spiral-bound brochure and portfolio (explaining our marketing strategy and design capabilities). The extraordinary format accomplished our main objective concerning this new client — it became a notable and well-received introduction.

1994 WORLD CUP POCKET GUIDE

This easy-to-use guide was designed for foreign and out-of-town World Cup Soccer sponsors and employees who needed basic information on details and schedules of the games. The stylized figure indicates an up-to-the-minute event and its camouflaged orientation appeals to a general audience. Diagrammed sports plays appear as a backdrop.

BLACK ENTERTAINMENT TELEVISION (BET) PROMOTIONS

BET, which heralded the era of dynamic cable programming and continues to broadcast a wide range of shows about timely topics, chose us to help them reach a variety of specific urban markets with creatively conceived projects.

The vitality of the updated design attracts a new viewing generation.

Bright colors and a collection of kinetic images work together to define the BET market, and are carried through on all components.

BET delivers... an exciting mix of entertainment, sports, music, news, public affairs programs and more...

BET delivers

national research

sports

sponsor

AFFILIATE KIT

With an already strong presence in the cable TV industry, BET sought to expand its affiliate base, strengthen existing markets, and broaden ad sales. The resulting kit design includes images that promote its youth, sports, news, and general programming on its demo tape, pocket folder, letter, and brochure.

BET ON JAZZ BROCHURE

The BET on Jazz channel, a 24-hour televised forum for jazz aficionados, spotlights its original programming with an uncrowded, easy-to-read layout and a whole rainbow of energetic colors. The montage style illustrates the distinction of this uniquely American musical genre.

BET ON WHEELS PROMOTION

We designed the marketing materials for BET's community outreach program, whose focal point was a portable concert stage where events were held that brought together a cross-section of urban audiences. Elements such as an unusual tab closure and a curved shape that echoes the promotion's "wheel" make the folder a stand-out.

design solution for this U.S. Postal Service project used a novel approach to eliminate the possibility of a stale, neutral-hued look.

THE SILENT SCREEN COMMEMORATIVES BY CHARLES CHAMPLIN

IN 1989, THE POSTAL SERVICE COMMISSIONED AL HIRSCHFELD, THE GREATEST AMERICAN CARICATURIST OF THIS CENTURY, TO DO A SERIES OF 20 SKETCHES OF GREAT PERSONALITIES FROM THE WORLD OF ENTERTAINMENT. ★ THE FIRST SET OF THESE SKETCHES, "THE COMEDIANS," WAS ISSUED IN 1991. IT INCLUDED THOSE UNIQUE AND IMMORTAL PERFORMERS JACK BENNY, EDGAR BERGEN AND CHARLIE MCCARTHY, FANNY BRICE, ABBOTT AND COSTELLO AND LAUREL AND HARDY. ★ THIS SECOND SET OF HIRSCHFELD SKETCHES CELEBRATES THE MAGICAL WORLD OF THE SILENT SCREEN — NINE ACTORS AND ACTRESSES AND ONE CLASSICALLY COMIC ENSEMBLE WHO MADE MAGIC IN THE DAYS BEFORE THE MOVIES LEARNED TO TALK AND SING. ★ FROM THEIR FLICKERING, ONE-REEL BEGINNINGS AT THE TURN OF THE CENTURY, MOVIES CAPTURED THE ATTENTION, AND THE AFFECTION, OF THE WORLD AS NO FORM OF ENTERTAINMENT HAD EVER DONE BEFORE. THE MEDIUM WAS STILL VERY YOUNG WHEN THE FIRST GREAT INTERNATIONAL STARS OF THE SILVER SCREEN APPEARED, INVADING THE DREAMS AND THE IMAGINATIONS, EVEN SHAPING THE ASPIRATIONS OF THOSE WHO WATCHED THEM. THEIR STARS WERE EXTRAORDINARILY BEAUTIFUL OR EXTRAORDINARILY HANDSOME, OFTEN EXOTIC AND MYSTERIOUS IN SILENT DAYS, JUST AS OFTEN VIRILE, CLEAN-CUT, HOMESPUN. THEY HAD NO VOICES

INTRODUCTION

Color was added with computer-generated tritones and duotones, and animation was implied with an inventive montage cover and layout. Related icons indicated paragraph breaks.

THE WASHINGTON POST HOTEL PROGRAM

To encourage local hotels to offer its daily newspaper to guests, the *Post* selected us to design an innovative promotion that would also reflect on the high standards of the participating hotels.

Sophisticated elements, including a box to house the materials and a tissue paper band securing the three brochures, upgraded this promotion so that it seemed to recipients as if they'd received a gift instead of a marketing campaign.

WARDS
CORPORATE IDENTITY

To communicate the philosophy of an organization supporting the professional care of animals used in research, the right proportions of ethos and professionalism were combined to develop a trademark that is thought-provoking, but not sentimental.

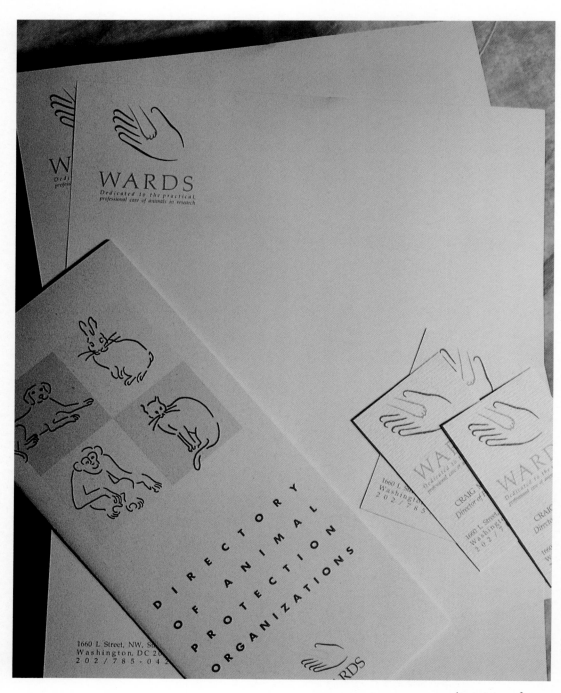

An austere and highly recognizable visual was the result of this approach, illustrating a caring relationship and response between humans and animals.

DEDICATED TO THE PRACTICAL
PROFESSIONAL CARE
OF ANIMALS IN RESEARCH

WARDS

GEORGE WASHINGTON UNIVERSITY VIEWBOOK

How do you appeal to college-bound, visually oriented youth and still hold the interest of their more sedate parents? Vibrant color choice, authoritative size, and powerful, attention-getting design gave this publication an aggressive, upbeat tone tailored to the theme of the George Washington University project: "Something Happens Here."

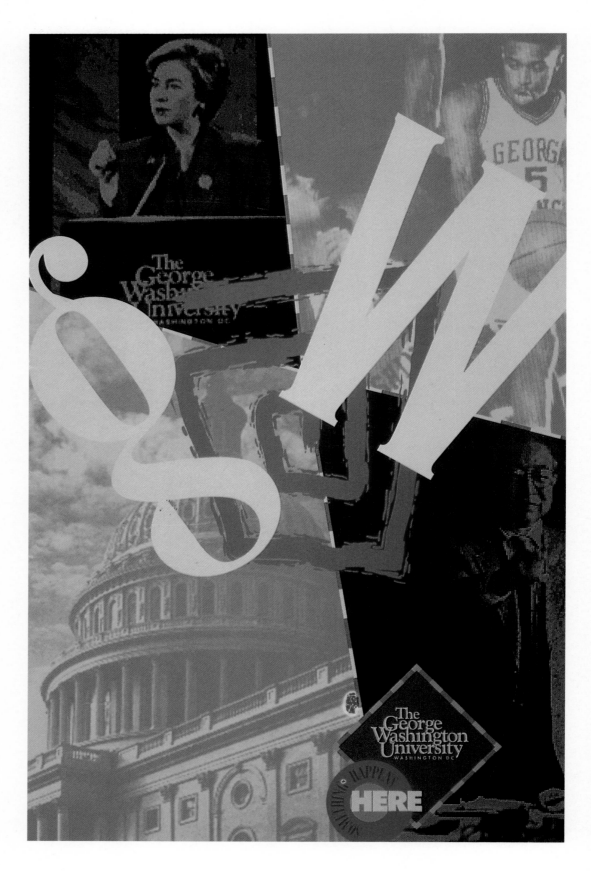

**VACATION TV
LOGO AND
PROMOTIONAL POSTER**

A television channel found in hotels throughout Asia, Vacation TV (a.k.a. The Vacation Channel until their name change) broadcasts tourist information about local and neighboring cities and countries. Its logo had to reproduce well, regardless of size, and transfer easily to different applications. Several approaches were undertaken to best capture the festive mood and the idea of "vacation" in an identifying symbol, with the preferred image the one most indicative of excitement, intrigue, and timely topics.

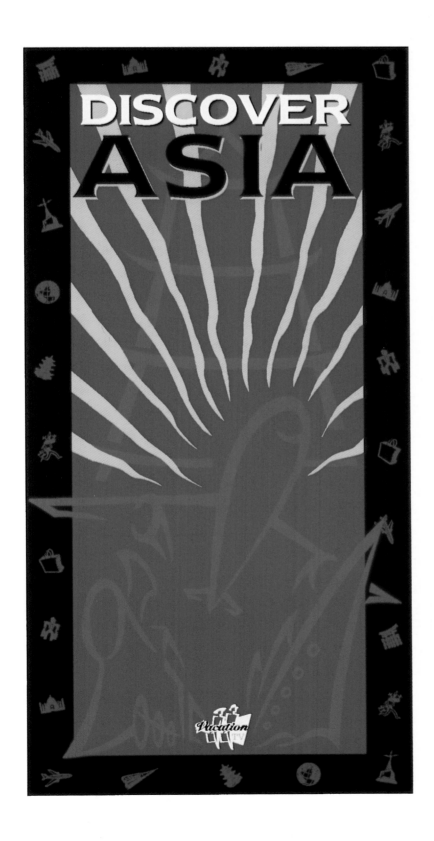

"SHARK SAFARI" P.O.P. DISPLAY

The Discovery Channel's popular television series dedicated to sharks needed a dynamic display that would promote both the program and the related sweepstakes. This contemporary design worked well using bold typography to catch the attention of passers-by, and colorful zig-zags around the fish to make it look less frightening, broadening the sweepstake's appeal.

KI RESEARCH CORPORATE IDENTITY CAMPAIGN

To downplay the common "slick" look that has become a marketing image cliché for many computer companies, these direct mail pieces and product description brochure for kiNet software emphasized the elegance of brushstrokes, giving the logo a softer look which harmonized with its tasteful elements.

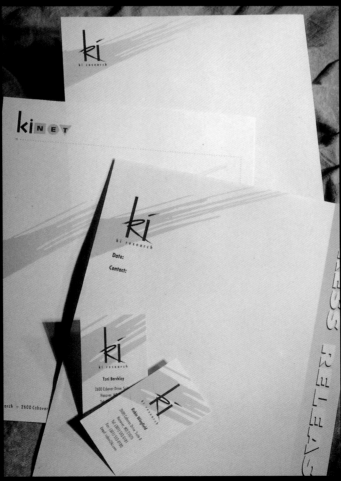

MONTAGE CORPORATE IDENTITY AND PRODUCTS

To feature their line of business papers, the clients wanted a catalog that would stand out immediately in a pile of mail, with a cover that would entice the reader to open the book. The all-purpose storage box modified the design features and colors from the corporate letterhead.

Similar to the client's paper styles, which can be effectively mixed and matched, the "Montage" masthead used different fonts to show how contrasting graphic elements — demonstrating both variety and consistency — can work well together.

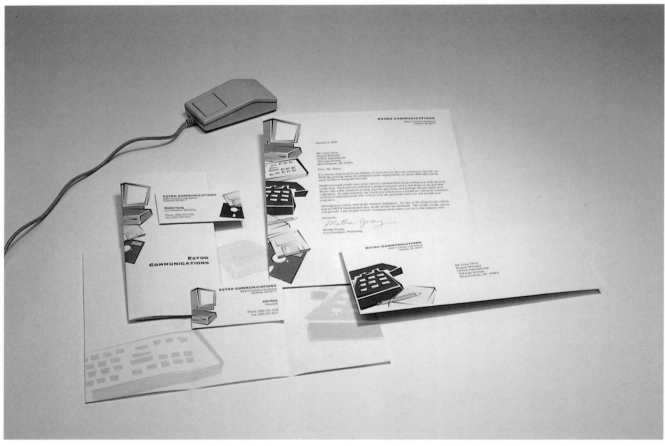

Two of the several lines of stationery, business cards, and brochures demonstrate contrasting styles to address the preferences of a wide range of companies. Folders house an assortment of paper designs that can be used on laser printers.

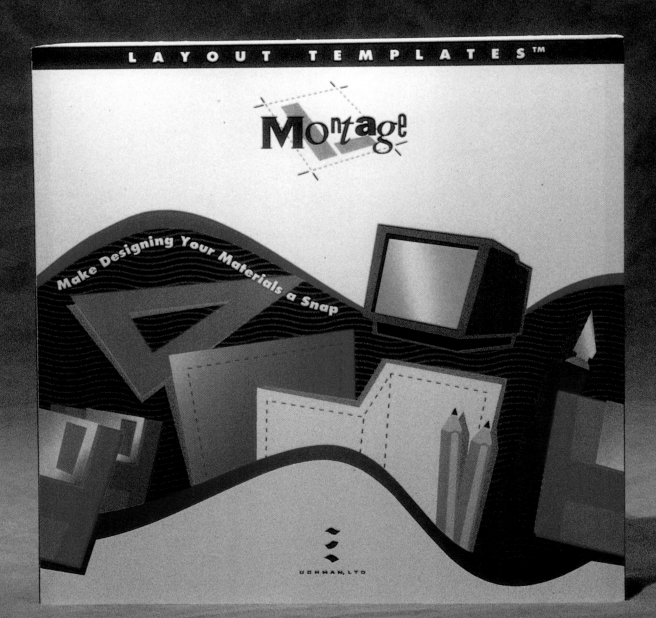

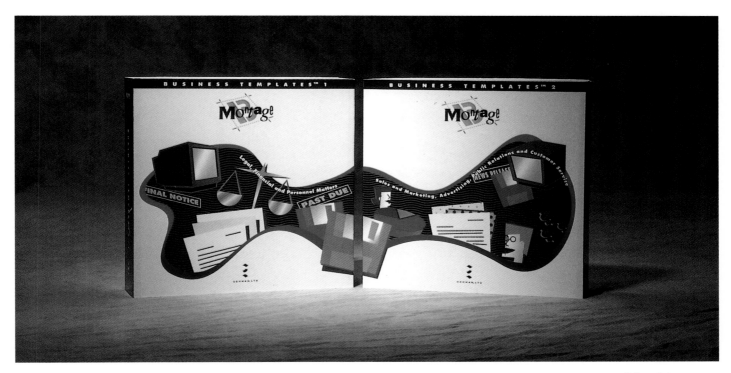

Although images displayed on the front of the package portray distinctive and separate subjects, their "connection" demonstrates the relationship between software templates (for brochures, business cards, and letters) and software concentrations (legal/accounting or PR/advertising). A "digital" illustrative style underscores the technological theme.

ADONIS SPRING WATER

Grand Palace Foods International of Thailand distributes a line of food products targeting Westerners traveling or living in Asia. With its delicate, virginal illustration, the label for Adonis Spring Water suggests freshness and purity.

IROQUOIS SPRING WATER

Sold on reservations of the Iroquois Nation, this spring water's identity features bold graphics in earthy colors reminiscent of this culture's traditional art.

FOOD MARKETING INSTITUTE (FMI) CONFERENCE BROCHURE

The Advertising/Marketing Executive Conference is one of several annual events organized by FMI. On this brochure, graphic illustrations of healthy foods mingle with items typically found on the desk of a busy ad executive. Hints of verdant palm fronds subtly reveal the meeting's tropical Florida locale.

GRAPHICALLY PLAYFUL

WITH A TILT OF TYPE, A CONTORTION OF IMAGE, OR AN ACCENT OF COLOR, DESIGN CAN BE ANIMATED TO EVOKE THIS WHIMSICAL STYLE THAT DRAWS VIEWERS IN AND MAKES THEM FEEL WELCOME. Although it is especially appropriate for children and the young at heart, this style is not theirs exclusively. Many projects could benefit from a little lighthandedness, a sense of humor, or a parody of a more serious approach. Vagrancies in color and illustration can be factors in animated design; but breaking rules of normalcy with a smile can be enough. One may arrive at a playful vision by altering the traditional components with a bit of entertainment. It is never immature design, however, but it does seem to require a perspective from a less serious soul, one that can unearth the quirkiness residing beneath every sober sketch or grim graphic. Our playful projects allow us to delve into our caches of images that are a little more odd and eccentric — ideas that depend heavily on wit and simile and puns, and on our ability to make them work in a marketing environment.

NATIONAL PASTA ASSOCIATION NEWSLETTER

Full of four-color photographs and light-hearted illustrations spiced with wit, the design of *Positively Pasta!*, the quarterly newsletter of the National Pasta Association, is a welcome departure from most trade association publications.

"FILMS FROM THE FRINGE FOUR" POSTER

It's not only animated advertising, it's an original approach for an American Institute of Graphic Arts (AIGA) invitation. The design is built as a poster that echoes the classic movie poster style of early Hollywood, from typography to vivid color treatments.

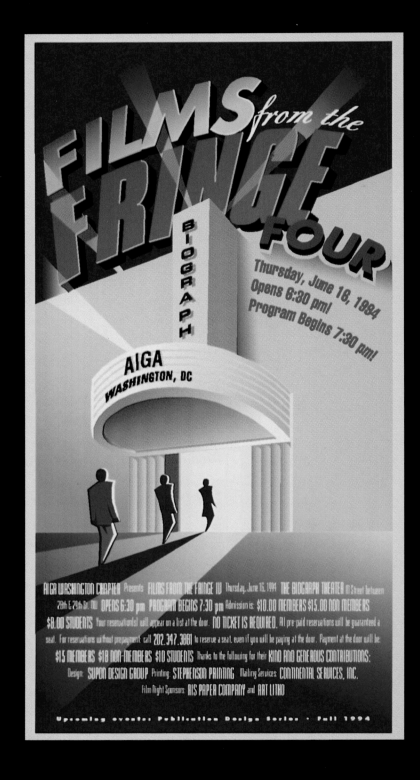

CORPORATE IDENTITY AND MARKETING PROMOTION

The Direct Marketing Association of Washington asked the studio to design a marketing campaign celebrating a Halloween party for a hypothetical client — Brimstone, a fraternity of witches and warlocks — that would demonstrate our range and creative skill.

Using an advertising parody, our designers created a humorous logo, campaign, and T-shirt premium for guests attending Brimstone's make-believe retreat.

TIME CAPSULE NOTECARDS

From Supon Design Group's product division, highly imaginative notecards organized in a wooden box illustrate a little boy's dreams 50 years ago of a future world.

The project demonstrates an innovative way to present a concept as packaging. It was so successful, it caught the interest of organizations like the Smithsonian Institution, which sells the product in its museum shops.

NINE CRITICAL QUESTIONS TO HELP YOU KNOW TV

1. Is this a documentary?
2. What is the filmmaker's purpose?
3. How does the filmmaker's purpose shape the content?
4. How do language, sound and images manipulate the message?
5. What techniques enhance the authenticity of the message?
6. What techniques enhance the authority of the message?
7. How do different viewers interpret the same message differently?
8. What techniques involve or engage the viewer in the message?
9. Who makes money from the message?

KNOW-TV IDENTITY

Know-TV is a series of workshops which promote the use of cable television in educators' curricula. Logo and identity applications shown here include a poster, manual, video box, and stationery.

PRODUCT DESIGN

A company with an interest in worldwide ecology and preservation, A Planet Called Earth asked us to develop original designs of animal representations for application on T-shirts and stationery. Graphic, abstract figures were developed so as not to show preference for any one species, and the aspect of Earth-friendliness was created by using visual puns — the turtle shell and fish bubbles become global metaphors.

WOLF TRAP CAMPAIGN

An outdoor theater for the performing arts located in a Washington, D.C. suburb, Wolf Trap has a long history of booking a wide variety of exceptional music, dance, and theater events. Their promotional campaign required the introduction of an original design that could be modified for use on T-shirts, P.O.P. displays, brochures, posters, or a bus sign.

Hues were chosen to convey spring and summer — the height of the Wolf Trap season. The subway poster used an orange-yellow shade and a picnic illustration to suggest an evening's entertainment that Wolf Trap has popularized around its community.

These two icons formed the foundation of the Wolf Trap summer season identity. As additional promotional applications were developed, the art was subtly altered and expanded.

We wanted the "Wolf Trap" name to be very large on the bus back design so that it could be read from across the street or while the bus was in motion, yet not garish, so it would easily be incorporated into the art. The result was an effective blend of typography, carefully chosen text, and recognizable images.

With a strong interest in the aesthetic, the client did not want to be represented by a "computer" look. Although the design ultimately used technology in its creation, the designer initially drew the art by hand, then scanned it into the computer and placed colors behind it to create the effect.

We turned some of the happy-go-lucky figures into spot icons and used them throughout the campaign.

The fresh, outdoor theme was also conveyed successfully in fabric. Colors and icons transferred effectively to embroidery and stood out against their stark backdrops.

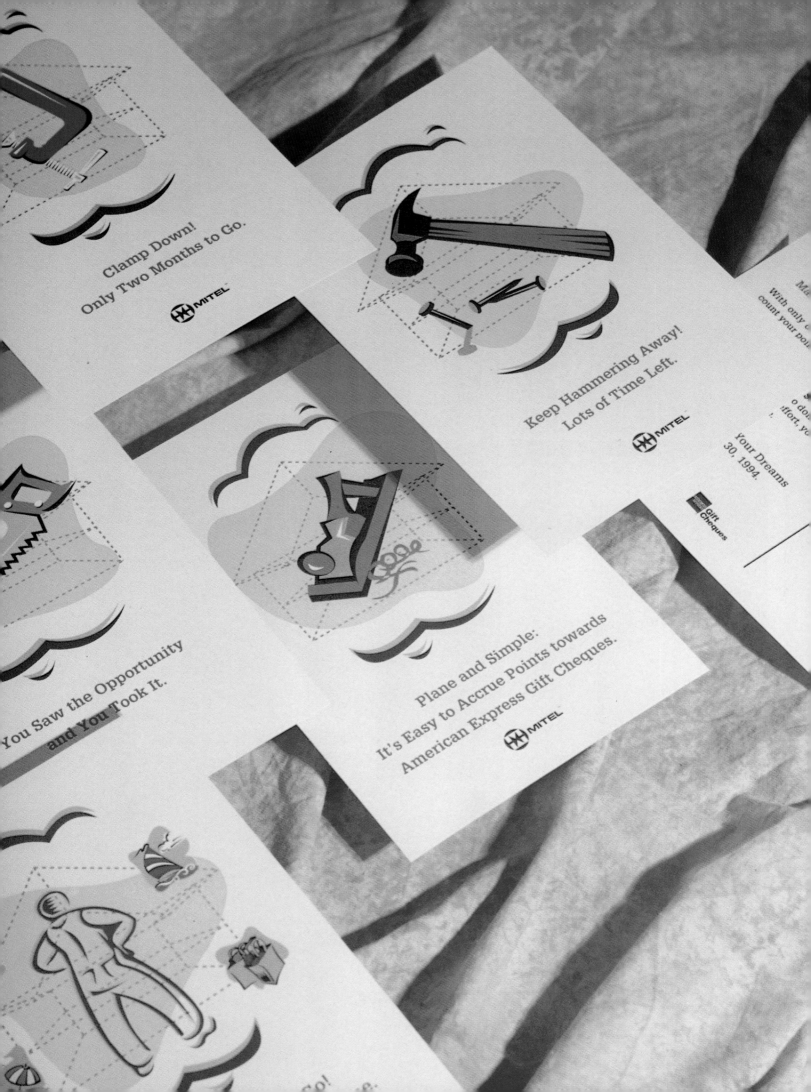

SALES INCENTIVE CAMPAIGN

A Canadian-based phone hardware manufacturer looking for a new way to promote a six-month sales incentive contest, Mitel was pleased with the "Building Your Dreams" slogan and the way we used construction imagery to create interest and also tell a story. Postcards mailed monthly to salespeople encouraged them to meet their goals.

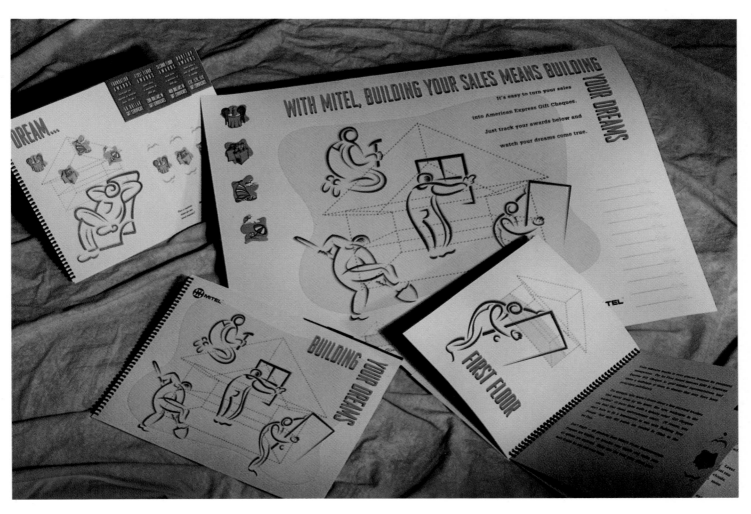

We stressed the "softer" sell and downplayed the typical "high-tech" look so prevalent for telecommunications firms. Printing the project on uncoated paper also provided contrast with the marketing materials commonly found in the industry.

NATIONAL TRUST FOR HISTORIC PRESERVATION CAMPAIGN

The lively and colorful poster heralds the annual event of an organization concerned with historical preservation and saving classic landmarks instead of replacing them with new construction. We designed the illustrative logo with the intent of applying color and making it into a poster that would promote the event. With its unusual perspective, it also attracts a younger generation to its cause.

It's your memory. I

PRESERVATION WEEK • MAY 8-14, 199

our history. It's worth saving.

NATIONAL TRUST FOR HISTORIC PRESERVATION

NATURALLY SPAGHETTI SAUCE

Its taste altered to appeal to the Asian palate, Naturally Spaghetti Sauce, distributed by Grand Palace Foods International of Thailand, still retains its homestyle identity. The label's woodcut-like illustration of a peasant hard at work reinforces this earthy image.

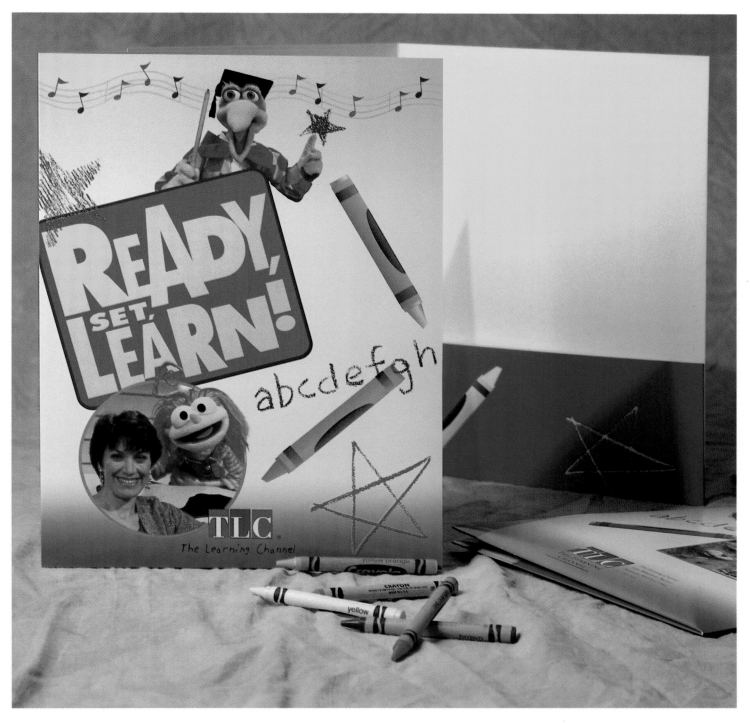

PROMOTIONAL POCKET FOLDER DESIGN

By using visuals familiar to preschoolers, this graphic solution for The Learning Channel (TLC) projected a friendy, accessible tone. From the choice of TLC's featured television characters, to the use of a youngster's childlike scrawl to adorn the covers, the design creatively and effectively addressed the company's objective: to distinguish these programs from their other shows for higher age groups.

NATIONAL WILDLIFE FEDERATION (NWF) FOREVER POSTER AND CALENDAR

This poster and calendar is part of NWF's Student Membership Program. Seven distinct images were merged to emphasize the interconnectedness of the Earth's varied regions, landscapes, and inhabitants.

Early sketches for NWF's Student Membership Program reveal our designers' diverse avenues of thinking. The final logo is shown far right.

ever

1 9 9 5

NATIONAL WILDLIFE FEDERATION
STUDENT MEMBERSHIP PROGRAM

Earth Day–Saturday, April 22

April 1 2 3 4 5 6 7 8 9 10 11 12 13 14 15 16 17 18 19 20 21 22 23 24 25 26 27 28 29 30

May 1 2 3 4 5 6 7 8 9 10 11 12 13 14 15 16 17 18 19 20 21 22 23 24 25 26 27 28 29 30 31

June 1 2 3 4 5 6 7 8 9 10 11 12 13 14 15 16 17 18 19 20 21 22 23 24 25 26 27 28 29 30

NATIONAL WILDLIFE FEDERATION

GRAPHICALLY
CORPORATE

WHETHER IT IS FOUNDED ON TRADITIONAL ATTITUDES OR ASSERTIVE MOTIFS, WE DISCOVERED EARLY ON THAT "CORPORATE" DOES NOT HAVE TO MEAN "BORING." Such visuals can achieve something dynamic by integrating active elements, yet maintain a sedate undertone by using purposeful ideas without flourishes or unneeded accessories. Corporate style is most suitable when it supports client confidence and underscores professionalism. Organizations who are custodians of great fortunes, have sobering missions, or tally long histories of global impact can often profit from a corporate style of design to represent them. Some businesses may do well in playing down a frivolous image, but their identities never need to match each other, or become so sedate as to be dull. This is an adaptable style that speaks easily to those at all rungs of the corporate ladder. This style bends the rules of tradition to produce work that brightens the company image through color usage and element relationship, and is always emboldened by a few rigid design basics to ensure a professional look.

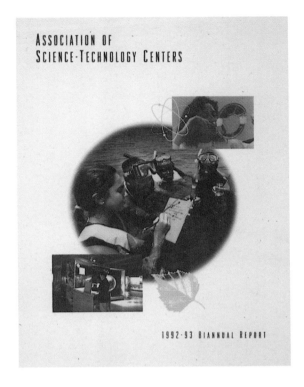

ASSOCIATION OF SCIENCE-TECHNOLOGY CENTERS (ASTC) BIANNUAL REPORT

The objective of this two-color biannual report is to present ASTC as a credible player in the fields of science and technology. With a very clean and strict grid, it summarizes the association's numerous educational and research programs, publications, and goals.

LIFE TECHNOLOGIES, INC. (LTI) CORPORATE BROCHURE

The challenge was to design a brochure where every spread contained 10 languages so that it could be understood by LTI's 1,400 international employees. A principle-directed biotechnology firm, LTI's brochure represents the guiding beliefs — their "ethos" — upon which the firm based its professional success. Subtle tints and appropriate photography communicated the seriousness with which LTI approaches its objectives.

LIFE TECHNOLOGIES™

WAY

The way we want to be

NOWATORSTWO

Nowatorstwo ma kluczowe znaczenie dla naszego powodzenia. Będziemy nowatorscy w opracowywaniu i wprowadzaniu na rynek produktów i usług, w polepszaniu jakości naszej działalności, w naszej współpracy z innymi, oraz w sposobach dalszego rozwijania naszej wiedzy zespołowej i wzrostu.

革新性について

革新的であるということが我々が成功するのに重要です。LTIは全てにおいて革新的であることを目指します。即ち、製品とサービスの開発と商品化に於いて、仕事の改良に於いて、パートナーシップの形成と維持に於いて、更なる組織としての学習と成長の為にとる手段に於いて革新的であることを目指します。

**NATIONAL COMMUNITY
AIDS PARTNERSHIP
CORPORATE IDENTITY**

Our task was to update the company's identity, using the existing type. These print pieces demonstrate appropriate design that not only deals with difficult issues, but has the flexibility to be used in many different applications, including a newsletter and annual report.

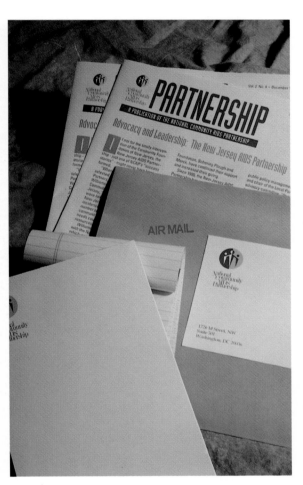

The logo's three figures link together community and individuals, and use of two colors disallows racial bias. The supporting pieces base their structure on that of the corporate identity.

INTERNATIONAL FOOD POLICY RESEARCH INSTITUTE (IFPRI) BROCHURES

A project for the IFPRI involved designing two brochures, one corporate and one describing their effort in Africa. Because of the market, we did not want an extravagant, expensive appearance; instead, we used uncoated, recycled paper and updated their traditionally conservative type treatment with a more candid and interesting approach. To enliven the Africa brochure, we avoided stereotypical images, choosing to incorporate a color pallette and ethnic motifs that portrayed the culture.

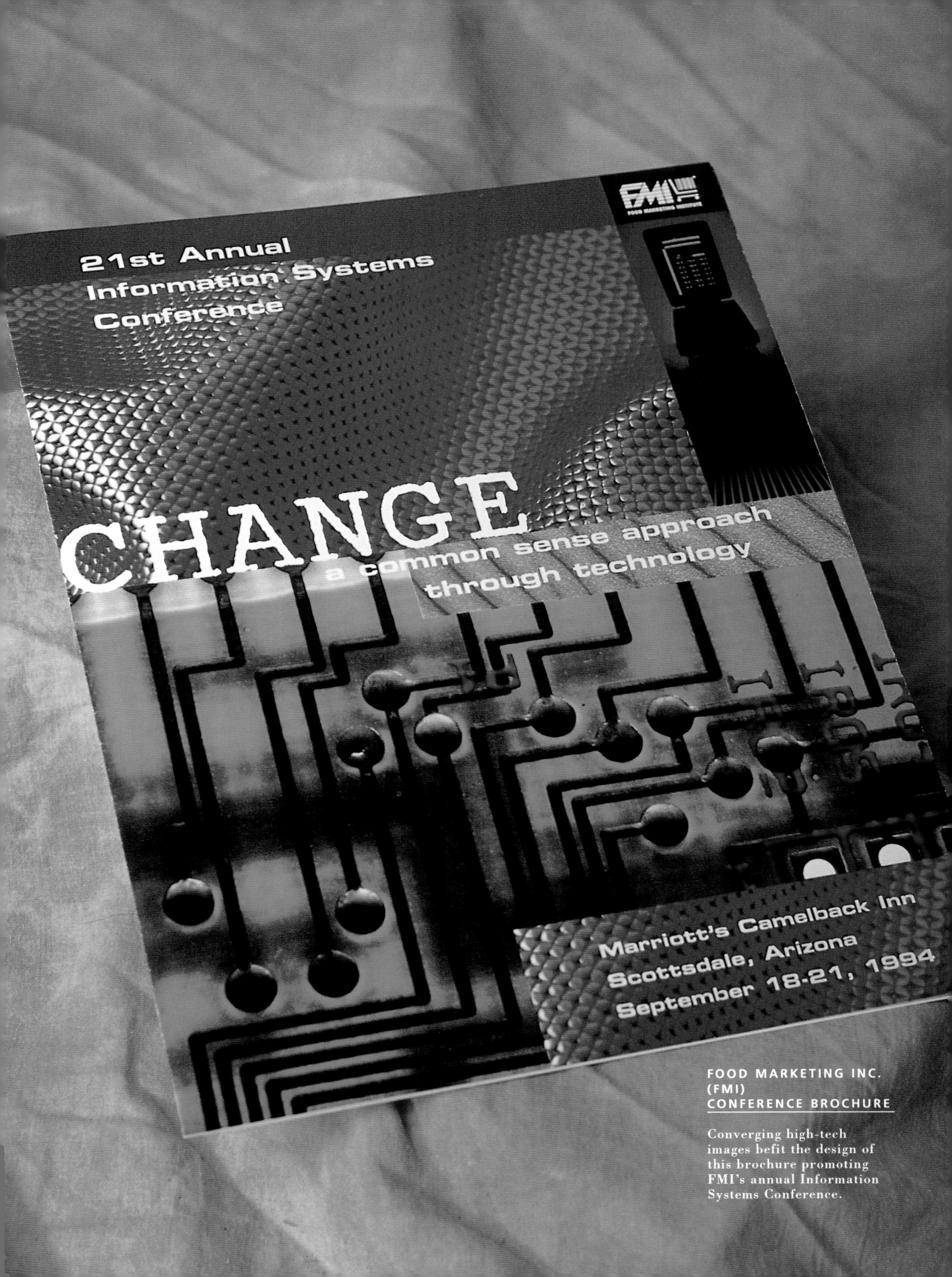

CHANGE
a common sense approach through technology

21st Annual Information Systems Conference

Marriott's Camelback Inn
Scottsdale, Arizona
September 18-21, 1994

FOOD MARKETING INC. (FMI) CONFERENCE BROCHURE

Converging high-tech images befit the design of this brochure promoting FMI's annual Information Systems Conference.

YOUR CHOICE TV
CORPORATE IDENTITY

The project's objective was to animate and enliven an existing logo for Your Choice TV, a groundbreaking television service from which viewers may eventually order programs. Letterhead, brochure, and a direct mail piece used a color palette similar to the existing design to associate the pieces, while keeping the professionally conservative tone.

To convey a highly technological feel, lines were incorporated around the letterhead logo to suggest a television screen.

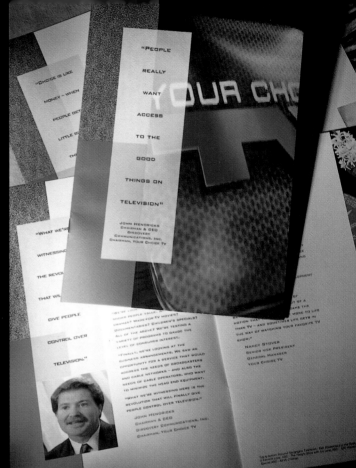

The corporate brochure cover made good use of texture and dimension to project movement and call attention to the CEO's quote. Inside, testimonials tell the story behind Your Choice TV as well as the objectives of the service, and scattered photos introduce the main players behind the effort.

FIFA CLUB LOGO

The French acronym for the Fédération Internationale de Football Association, FIFA is soccer's highest administrative authority. FIFA Club is a preeminent faction of members and World Cup sponsors' top executives. Using the existing FIFA logo as its foundation, the FIFA Club logo adorned member giveaway premiums.

FIFA CLUB

The logo's bold typography and clean, uncluttered look projects sedate professionalism while maintaining both the FIFA style and enough flexibility for use in numerous applications.

TIME-LIFE VIDEO PACKAGING

For a broadcasting series on archaeology, our designers developed an approach that looked bold on a shelf and stayed consistent through the series, while distinguishing each program. The Egyptian hieroglyphics, used throughout the series because they were so visually striking, were accented with blue tones instead of the traditional sand. The cover photograph and title hue — which coordinated with the photo colors — were the components that defined each package.

"NASDAQ CHALLENGE" PACKAGING DESIGN

Software that helps simplify the stock market — NASDAQ's idea for a giveaway to potential and existing clients had outstanding appeal. The packaging included a disk and a "NASDAQ Challenge" quiz that provided instant results to test takers who could win "money" with smart investment strategies. The typeface suggests a business-like tone and the dark colors imply strategy and intellect.

...up Health Association of ...rica knew their magazine, ...O, could use an updated ...gn — one that would be ...ing to the reader and ...ure varied layout, more ..., and different styles. ... approach integrated ...tration with type and ...red an assortment of ...ions in every issue, from ...ype to heavy illustration ...balanced combination of ...wo. The table of contents ...sign rendered it more playful.

THE Mind BODY connection

By Daphne B. White

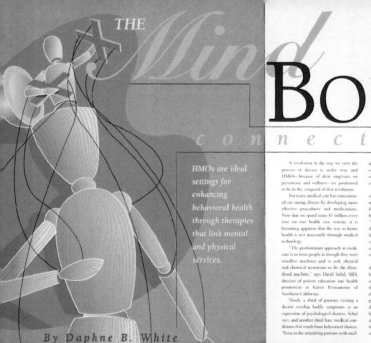

HMOs are ideal settings for enhancing behavioral health through therapies that link mental and physical services.

A revolution in the way we view the process of disease is under way, and HMOs—because of their emphasis on prevention and wellness—are positioned to be in the vanguard of this revolution.

For years, medical care has concentrated on curing disease by developing more effective procedures and medications. Now that we spend some $1 trillion every year on our health care system, it is becoming apparent that the way to better health is not necessarily through medical technology.

"The predominant approach in medicine is to treat people as though they were mindless machines and to seek physical and chemical treatments to fix the disordered machine," says David Sobel, MD, director of patient education and health promotion at Kaiser Permanente of Northern California.

Nearly a third of patients visiting a doctor develop bodily symptoms as an expression of psychological distress, Sobel says, and another third have medical conditions that result from behavioral choices. "Even in the remaining patients with medical diseases such as arthritis, heart failure, or pneumonia, the course of their illness is often strongly influenced by their mood, coping skills, and social support," he adds.

This critical mismatch between the real health needs of people and the usual medical response leads to frustration, ineffectiveness, and a gross waste of vital health care resources," concludes Sobel. "This approach ignores the striking fact that what goes on in our heads—our thoughts, feelings, and moods—can have a dramatic effect on the onset of some diseases, the course of many, and the management of nearly all."

Since 60 percent of all office visits are due to stress-related ailments that have both a physical and psychological component, a number of HMOs are now taking the lead in developing integrated services that meet both the psychological and physical needs of their members. Not only does such an approach improve patients' health, but many of these services have also demonstrated considerable potential for cost savings. Because they are structured managed care systems, HMOs are ideally suited to offer integrated health care services, says Jessie Gruman, executive director of the Center for the Advancement of Health (CAH), a Washington, D.C.-based health policy organization. Unlike fee-for-service plans—where primary care providers, medical specialists, and mental health providers are not completely integrated and are financially independent of each other—HMOs can offer patients "a real continuity of care," Gruman says.

"The word maintenance in 'health maintenance organization' means more than disease prevention. It means treating the patient as a whole person, and giving them new ways to solve old problems," Gruman says. "HMOs can offer a continuum of care over time, and a continuous between physical and mental health services."

A Strong Mind-Body Connection

Michael Von Korff, PhD, associate director of the Center for Health Studies at the Group Health Cooperative (GHC) of Puget Sound, has researched the management of depression and chronic pain in primary care settings.

"What we're realizing now is that any illness has both a physical and psychological component. That's true for back pain, diabetes, depression, or irritable bowel syndrome," says Von Korff. "There is a very complicated interactive relationship between the brain and the body."

Marketplace Survival: The New Rules

By Elaine Zablocki

Today's competitiveness means setting up systems that emphasize quality, local delivery, and product flexibility.

Given today's fast changing, competitive marketplace, meeting consumer needs and beating legislation that threatens the quality components inherent in well-integrated managed systems will pose new challenges as HMOs continue to evolve in the marketplace.

HMO Magazine asked a wide range of industry leaders to name the top areas where HMOs can be best positioned to meet these challenges in the coming decade. The top of their lists: accountability through quality performance; market and product flexibility; legislative might; and strong provider relations at the local level.

A New Accountability

Almost everyone we consulted cited increased pressures for producing measures of quality as the top issue HMOs face today.

As Barbara Hill, Prudential's vice president for health care policy, notes, "quality and cost go hand-in-hand. But all the focus on controlling costs is for CIGNA, says, "health care is becoming a business much more aligned with the rest of the economy. Purchasers, whether individuals or companies, don't want to spend large sums of money without quality and cost documentation. In the past health care has been sold primarily on reputation, but today, through quality measures such as HEDIS [the Health Plan Employer Data and Information Set], it is possible to offer documented performance." FHEDIS is one set of health plan performance measurements by which employers and other health care purchasers can compare health plan value (see HMO Magazine, March/April 1994, pp. 26-32).

Richard Leveles, executive vice president of Cypress, Calif.-based PacifiCare Health Systems, predicts "the plans that will be successful in the future will be those which acknowledge they are accountable for both cost and quality, and can demonstrate that they're doing a good job."

However, proving value through measures of quality demands a long-term investment in new data collection systems, ments today as a matter of daily business. They have to do a thorough chart review or take samples or do exhaustive tape reviews to prepare the measurements. So any necessary step for the future will be information systems that produce quality measures efficiently.

Hill notes that plans today have to prepare quality data in varying formats to meet requirements from different purchasers. "We ought to have one accreditation agency for all HMO payers," she says.

In addition, the calls for uniform quality standards. "There should be one standard that is acceptable and fair for all payers and all patient populations. For example, I don't think that Medicaid patients and commercially insured patients should have mammograms at different ages. There should be one standard for everybody. Then all accountable health plans should be measured against the standards, so consumers can easily compare Plan A with Plan B."

Karen Ignagni, GHAA President and CEO, notes, "we have been working very hard to ensure that the rules of the game will be clearly set out, and won't vary sub-

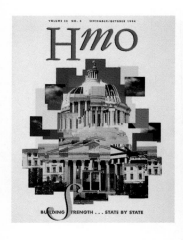

HEALTH CARE INTEGRATION: A GREATER WHOLE

TWO MEDICAL ORGANIZATIONS COMBINE TO FACE A COMPLEX AND COMPETITIVE MARKET.

IN 1985, HEALTH CARE VISIONARIES PAUL ELLWOOD AND LELAND KAISER PRESENTED WHAT THEY PREDICTED WOULD BE THE MEDICAL ORGANIZATION OF THE FUTURE TO A MEETING OF THE AMERICAN GROUP PRACTICE ASSOCIATION (AGPA). THEY SPOKE OF VERTICALLY INTEGRATED HEALTH CARE ALLIANCES BETWEEN PLANS AND PROVIDERS—ALL UNITED TO IMPROVE QUALITY, CUT WASTE, CONTROL COSTS, AND PROMOTE UNIVERSAL ACCESS TO CARE.

BY PAUL A. SOMMERS AND ERIC P. SOMMERS

PRACTICE GUIDELINES

FROM THEORY TO 'BEST PRACTICE'

THREE CASE STUDIES ILLUSTRATE HMOs' EXPERIENCE IN DEVELOPING AND USING PRACTICE GUIDELINES.

By Elaine Zablocki

Today, as health care leaders attempt to improve clinical outcomes and limit costs, they increasingly turn to clinical practice guidelines to decrease variation in physician practice patterns. Successful guideline implementation can lead to improved use of preventive screening tests and immunizations, fewer C-sections, and better control of chronic diseases like asthma and diabetes.

PHARMACEUTICAL INTEGRATION:

A recent wave of mergers and alliances with PBMs is focusing the drug industry on outcomes research and disease management. How will it affect HMOs?

GOING VERTICAL

By Carl E. Peterson

Outcomes Research in HMOs:
Studies in Quality

By Judy Packer-Tursman

On the forefront of outcomes research, HMOs aim to determine the best approaches to health care delivery.

Throughout HMO magazine, illustrations became elements of the stories, not just pictures of them. Drop caps and bold headlines activated the page. A rigid three-column grid maintained the professional tone, although we removed the extraneous rules.

GRAPHICALLY
PERSONALIZED

TO SHRINK ONE'S PHILOSOPHY, OBJECTIVES, AND PURPOSE TO A REPRESENTATIVE SYMBOL IS A VISUAL CHALLENGE UNMATCHED ANYWHERE ELSE IN COMMERCIAL ART. The creation of an appropriate trademark involves the production of numerous sketches, erasures, and a singular ability to communicate paragraphs of information from a precise, uncluttered image. For designers, it becomes important to maintain a slightly altered perspective, literally seeing things in an entirely new way. Creating corporate marks means revealing invisible elements and structuring one idea so that it is strong enough to carry the identity. Designers can present a company with a look that will buoy it up from the recesses of the ordinary and suddenly make it a powerful force able to persuade people to buy its product or service or buy into its beliefs. Trademarks can be masterworks of summarization, and the examples we present here have confronted the challenge to produce ideas that are succinct and dramatic.

These logos each project orientations that are highly stylized, but distinctive to each organization.

From top:
The logo for the **V/S SPA AND MASSAGE CLUB**, a Thai health environment, embodies an image of purity emerging from clear and gentle lines and thoughtful symmetry. **A PLANET CALLED EARTH** tries to incorporate its environmental awareness into all its products. This image capitalizes on Earth tones and assertive elements to appeal to the urbane consumer. Sponsored by the National Trust for Historic Preservation, the **GREAT AMERICAN MAIN STREET AWARD** sports a logo which captures the simplicity of small town life and transforms it into a smart vision with bold lines and purposeful twists and angles.

The homogeneity of corporate identities that produce an undeniably diligent and enterprising look is demonstrated in this variety of symbols.

From top:
A progressive but composed look was the intent for the identity for **ACADEMY FOR EDUCATIONAL DEVELOPMENT**. The stroke and orb portray a forward-thinking firm with a worldwide interest. **CITY OPTICAL** concentrates on the spherical qualities of the "C" and "O" to produce a unified look that also appropriately resembles a pair of spectacles. A company that conducts billing for telecommunications firms, **COMMSYS** needed a look that introduced it into the technological field with dash and professionalism. The resulting identity used bold type and an appropriate symbol to create uniformity and earnestness.

AXENT, a firm specializing in software for computer systems security, requested a symbol that would act as a definition of their company separate from their name. The swash that comprises the crossbar of the "A" portrays the contemporary flair for which the company is recognized.

Above:
BKK RADIO, with headquarters in Asia, broadcasts music that appeals to young audiences, and the trademark's jaunty angles and animated qualities creatively illustrate their contemporary philosophy.

Facing page:
Representing a wholly updated and streamlined interpetation of the zodiac symbols, these icons were used in the horoscope column of the Asian magazine *MEDIA DELITE*. They were created by hand on scratchboard.

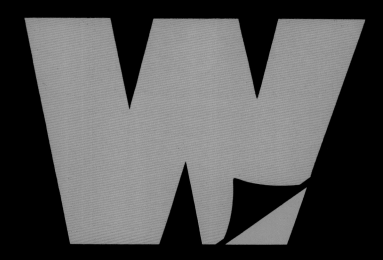

Here you'll find typography representations that don't just sit idly on the page, but actively define corporate services, products, and entities with imaginative interpretations of the alphabet.

Above:
WALSH WALLPAPER SERVICES, a wallpaper installation firm.

Facing page,
clockwise from upper left:
ADVENTIST HEALTHCARE MID-ATLANTIC, a family of affiliated hospitals; Dr. Steven Bunn, dentist; **CHARLES BUTTON COMPANY,** button manufacturer; **HARRIS CHAIR COMPANY,** chair manufacturer; **ROBINSON RADIOLOGY,** radiologist; and **ULLMAN PAPER BAG COMPANY,** bag manufacturer.

MEE DERBY & COMPANY is an employment placement agency headquartered in Washington, D.C. The three-figured design of the logo echoes the professional triad of employer, agency, and prospect. The primitive design style and amber and green colors suggest earthiness, solidity, and permanence—separating Mee Derby from temporary placement agencies.

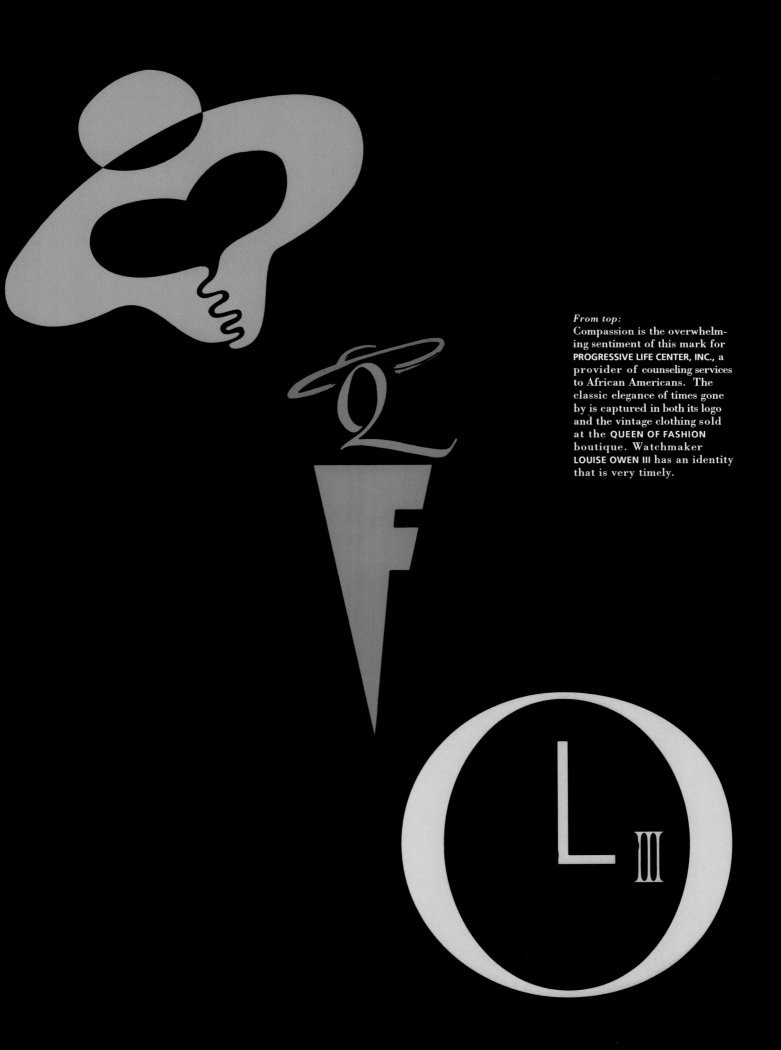

From top:
Compassion is the overwhelming sentiment of this mark for **PROGRESSIVE LIFE CENTER, INC.**, a provider of counseling services to African Americans. The classic elegance of times gone by is captured in both its logo and the vintage clothing sold at the **QUEEN OF FASHION** boutique. Watchmaker **LOUISE OWEN III** has an identity that is very timely.

**TELEVISION PROGRAM
DESIGN ICONS**

How could The Discovery Channel customize the brown paper bags used to carry its marketing kits without custom printing and expense? With only a few days' lead time, we developed a series of prominent program icons that were manufactured into rubber stamps, which are used to imprint the image of an appropriate program onto each bag before it is delivered to interested companies.

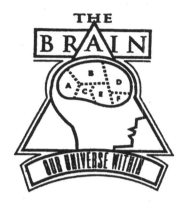

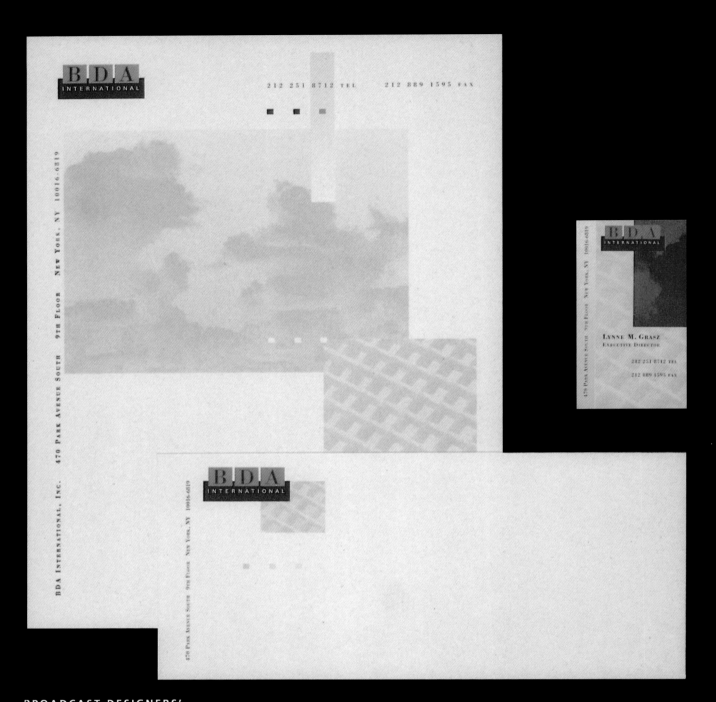

**BROADCAST DESIGNERS'
ASSOCIATION (BDA)
STATIONERY**

A collage of abstract, high-tech images suggests today's rapidly evolving broadcast industries.

**PARADISE WILD
IDENTITY**

Paradise Wild is a retail store which sells nature- and animal-related gift items. The store's identity, shown here on its stationery, was also applied to shopping bags, signage, and other items.

DIVERSE HANDEL IDENTITY

A retail shop selling a wide array of trendy, offbeat housewares, the Diverse Handel identity consists of a very contemporary rendering of a home interior.

Its initials set in stone and a sundial were two of our early concepts for the SAA identity.

SOCIETY OF AMERICAN ARCHAEOLOGY (SAA) CORPORATE IDENTITY

When developing a new identity for a conservative archaeological organization, the designers composed a solution that authentically reflected the science, but did not look old and tired itself. Instead, the cohesiveness of the concept was stressed, which fit a multitude of applications, from products to campaign materials to publications. The solution made use of generic, hieroglyphic-like symbols that suggested the ancient pictographic script but did not emphasize any particular culture or period, making the images timeless and widely applicable.

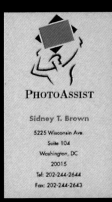

PHOTOASSIST IDENTITY

After long hours of searching, the individual on this identity for a photo research group could be proclaiming, "Eureka! I've found it."

GIGGLE IDENTITY

Giggle is a chain of international gift shops with locations in five different countries. Even though the letterhead is printed on both sides, the one-color design is very economical.